Great Events

The GUNPOWDER PLOT

Written and Illustrated by Gillian Clements

W
FRANKLIN WATTS
LONDON•SYDNEY

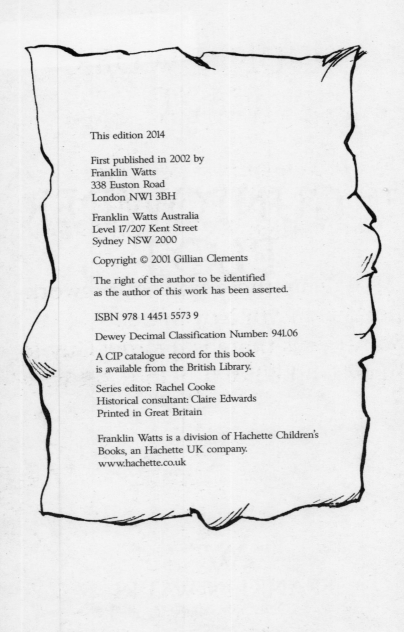

This edition 2014

First published in 2002 by
Franklin Watts
338 Euston Road
London NW1 3BH

Franklin Watts Australia
Level 17/207 Kent Street
Sydney NSW 2000

ISBN 978 1 4451 5573 9

Dewey Decimal Classification Number: 941.06

A CIP catalogue record for this book
is available from the British Library.

Series editor: Rachel Cooke
Historical consultant: Claire Edwards
Printed in Great Britain

Franklin Watts is a division of Hachette Children's
Books, an Hachette UK company.
www.hachette.co.uk

The GUNPOWDER PLOT

*"Remember, Remember
the Fifth of November,
Gunpowder, Treason and Plot."*

We recite these lines at firework
parties on 5th November.
Sometimes there is a straw Guy to
burn on the bonfire, too. But why?

Because, on 5th November 1605, a man called Guy Fawkes and his friends tried to blow up the House of Lords. They wanted to kill King James I.

Guy did not lead the gang. But he played a big part in what happened.

Guy Fawkes was born in April 1570, in York – a big city full of merchants and scholars. It had a very fine cathedral. Christian worshippers prayed there for Elizabeth, their queen. But not all Christians supported Elizabeth.

5

Christians in England were divided; the Church was split into two main parts.

On one side were the Roman Catholics. They believed the head of the Church was the Pope, who ruled from his great Vatican palace in Rome.

On the other side in England were the Protestants. Queen Elizabeth was their leader. She followed in the footsteps of her father, Henry VIII. He had argued with the Pope about the best way to worship God. He broke away from Rome and set up his own 'Church of England'.

In York – as in the rest of England – Catholics and Protestants lived side by side. But Catholics were afraid of the Queen. She sent them to prison, or fined them, if they did not agree with her Protestant beliefs.

Some priests on the run were executed... if they were found hiding in priest-holes.

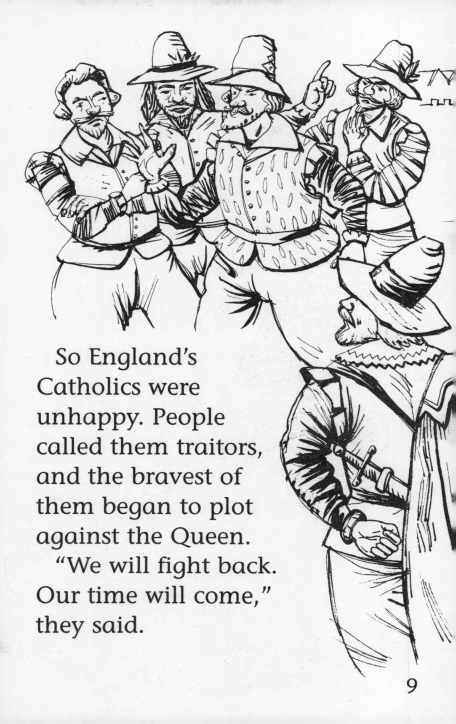

So England's Catholics were unhappy. People called them traitors, and the bravest of them began to plot against the Queen.

"We will fight back. Our time will come," they said.

But the Catholics' troubles continued. Things were so bad that in 1587 even Elizabeth I's cousin, the Catholic Mary Queen of Scots, was executed for treason.

I will finish the game and thrash the Spaniards too.

King Philip of Spain decided to help England's Catholics. He sent a huge Spanish Armada to England. But Elizabeth's navy, led by Sir Francis Drake, scattered the Spanish fleet.

Back in York, despite all these problems, Guy Fawkes made a decision. He would leave his dead father's Protestant religion. "I'll be a Catholic, like my mother and my friends," he said.

Guy Fawkes hoped that King Philip of Spain would try again to help put a Catholic on the English throne – so in 1593 Guy joined the Spanish army.

"If I learn about gunpowder, and how to kill rebellious Protestants ... these will be very useful skills!"

King Philip of Spain paid Guy Fawkes well for his work. And Guy became known as Guido.

Back in England there were plots
to kill the Queen. The famous
Earl of Essex was beheaded for
his treachery. John Wright, a
school friend of Guy Fawkes, had
plotted with Essex. He was lucky

to escape with his life.

"The Queen is dying, anyway,"
thought Wright. "The new King
has promised to be fair to
Protestants and
Catholics. We
must wait
and see."

In 1603 the old Queen died. "Long live the King!" the people cried, "God save King James I!"

King James was a Protestant, but English Catholics trusted this new king from Scotland.

"He is a wise man. And his mother was Mary Queen of Scots," they said. "A Catholic! He will be fair to us."

But James did not keep his word. "Our King has betrayed us," said the Catholics. One of them, Robert Catesby, was so angry that he started a plot against the King – he had plotted against the old queen too. He asked his friends, Yorkshiremen Thomas Percy and John Wright, to join him.

"BY HEAVEN WE'LL HATCH ANOTHER PLOT," cried Catesby. "This time we'll KILL the King. And I'll pick our best, truest Catholics to help."

Catesby thought for a moment. "But who?"

He wrote down three names.
There was John Wright's brother,
Kit, Thomas Wintour, a scholar,
traveller and soldier – and
finally, John Wright's old
schoolfellow, Guy Fawkes.

"We could use a man
who works with
gunpowder,"
grinned
Catesby.

Thomas Wintour sailed to Flanders to speak to Guy Fawkes. "Guido, come home. Join our fight against the King."

In April 1604, they returned to England together.

20th May
In the year of
Our Lord 1604.
Make haste
to the Dog &
Duck!

In the dark tavern, the plotters sat nervously. Catesby beckoned them over to discuss the plot.

"Guard against spies, my friends. For soon we will kill the King. When he opens Parliament, we will BLOW him to pieces. And his Protestant Lords with him!"

"But how?" asked Thomas Percy.

"Why, we will put gunpowder under the House of Lords, and light the fuse," Catesby replied.

"And then?"

"Then Catholics will rule this land."

"I've rented a house. It's by the House of Lords," Catesby added. "Guy Fawkes, you will be our 'caretaker'. We will dig in the cellar, and tunnel from the house to below their Lordships' Chamber. Then our caretaker will light the gunpowder, and..."

They could imagine the panic.

I've rented a house...

"The gunpowder must be kept safe till we use it," Guy Fawkes said. Catesby made Robert Keyes its guard. The gang did not see a shadowy figure slip away from the tavern.

Was there really a tunnel? No one really knows. It would be difficult to keep such a secret. No one heard any digging...

3

A few months later Catesby's men met again – at the theatre.

"Good fortune is smiling upon us – I can scarcely believe it," whispered Catesby. "The coal cellar under the House of Lords is empty! The merchants have gone and it's ours to rent!"

Meanwhile, the gang grew to thirteen.

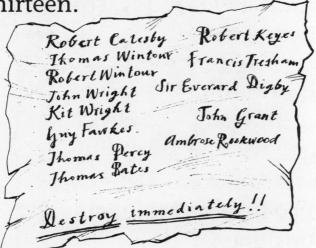

"Can so many keep our plan secret?" thought Guy Fawkes. "And now there's the plague. King James won't open Parliament this year."

It was 1605, late October. The gunpowder was ready, safe in Lambeth.

Across London, a letter was brought to Lord Monteagle. It was unsigned.

To the ryght honorable the Lord Monteagle

The letter warned Monteagle to stay away from the opening of Parliament. Monteagle was a Catholic, but loyal to the King. He took the letter to Robert Cecil, the King's most important adviser!

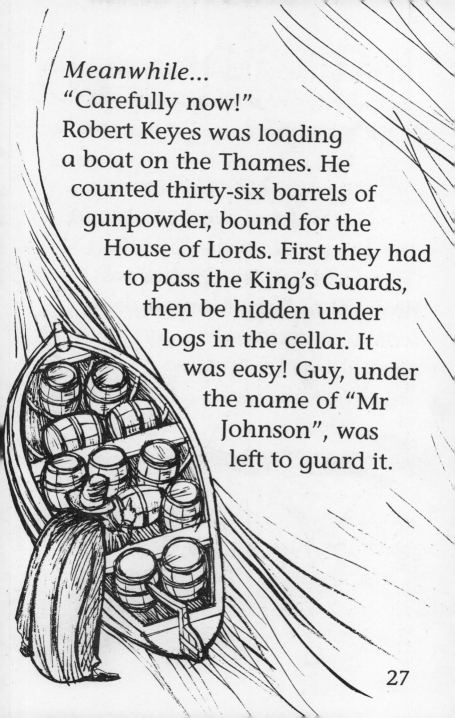

Meanwhile...
"Carefully now!"
Robert Keyes was loading
a boat on the Thames. He
counted thirty-six barrels of
gunpowder, bound for the
House of Lords. First they had
to pass the King's Guards,
then be hidden under
logs in the cellar. It
was easy! Guy, under
the name of "Mr
Johnson", was
left to guard it.

On 4th November
1605, there were shouts in
the House of Lords. "SEARCH
THE CELLAR, BY ORDER OF
THE KING," cried a guard.
"We're betrayed!" Guy
thought as he scuttled
into the shadows.

At much the same time, King James was reading Monteagle's letter again.

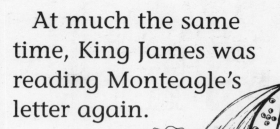

My Lord out of the love i beare to some of youer frends I have a caer of youer preservation.

26 th October in the year of our Lord 1605. Lord

Monteagle. Avoid attending Parliament. Those that do shall receive a terrible blow.

"It seems they mean violence, my Lord Cecil," he said.

"I think you may be right, Your Majesty," replied his adviser.

29

In the cellar, Guy watched the noisy guards leave.

"Nothing here!" they shouted. "Just that tall and desperate fellow. A servant I'd say."

Guy thanked God for his lucky escape, and he stayed out of sight.

Just a tall and desperate fellow.

Who wrote to Lord Monteagle? Was it one of the plotters? Or a government spy, who had watched them for months?

Why hadn't Robert Cecil done anything about the letter earlier? Did he hope to catch more of the gang? It seems he and the King had known about the plot all along – but now it was time to act.

It was just before midnight on 4th November.

In the cold, dark cellar, lantern-light picked out the scuttling rats.

Very soon King James's Parliament would meet above the deadly gunpowder store.

Guy was ready to light the fuse – his role in the plot.

"When I light this, the King will be blown to bits!" he laughed. "England will be Catholic in a blink of an eye."

"BANG!!"

The doors burst open!

Guy panicked. "Who told the Guards?"

In a muddle of footsteps and shouts, they grabbed him. Soldiers rushed straight to the woodpile.

"GUNPOWDER!" they exclaimed.

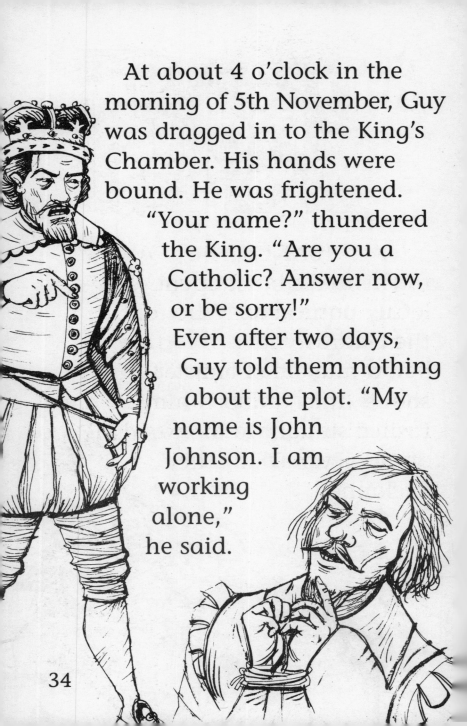

At about 4 o'clock in the morning of 5th November, Guy was dragged in to the King's Chamber. His hands were bound. He was frightened.

"Your name?" thundered the King. "Are you a Catholic? Answer now, or be sorry!"

Even after two days, Guy told them nothing about the plot. "My name is John Johnson. I am working alone," he said.

"TAKE HIM TO THE TOWER!" raged the King. "The rack will loosen his tongue!"

It took two days of torture before Guy cried out, "STOP NOW! I am Guido Fawkes. I will give you the names of my friends."

The names of other
Principal persons that were made
privy afterwards to this horrible
conspiracy.
Everard Digby (Knight)
Ambrose Rookwood
Francis Tresham   (signed)
John Grant   Guido Fawkes
Robert Keyes Wintour   Guido

"Find these men, and bring them to the Tower!" the King ordered his guards.

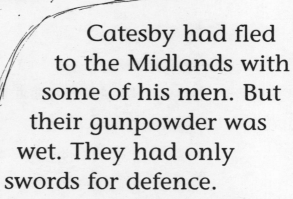

Catesby had fled to the Midlands with some of his men. But their gunpowder was wet. They had only swords for defence.

As the King's men surrounded their house, Catesby shouted, "We'll make our stand here! Dry the powder, and prepare to fight!"

"BOOM!!" Gunpowder
exploded by an open fire,
blinding John Grant.

38

"OPEN FIRE!" Musket-fire whistled and thudded into Holbeach House, and the Wrights died. A single bullet killed Catesby and Percy.

Only eight survivors were
brought to Westminster Hall, for
a very short trial. Finally, the
judge spoke.

"We have found you guilty of treason. You will suffer a traitor's death. At your place of execution you will be hung by the neck, taken down alive, and cut to pieces."

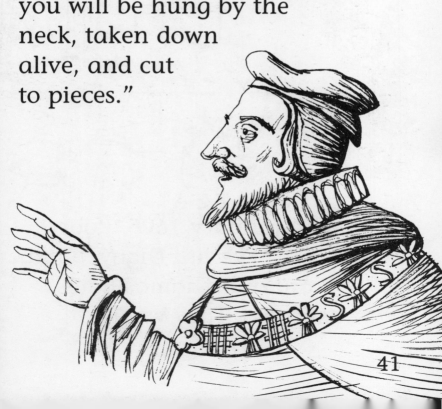

On 29th January 1606, Robert
Wintour, Sir Everard Digby, blind
John Grant and Thomas Bates
were executed. They were

followed the next day by Thomas
Wintour, Ambrose Rookwood,
Robert Keyes and, last of all, Guy
Fawkes – the plotter we remember.

Their heads were stuck on
spikes, and birds pecked at their
dead faces.

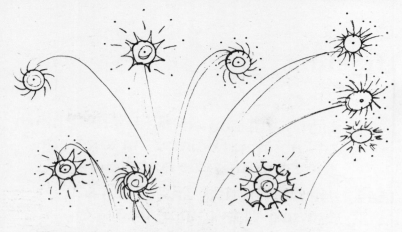

In the end King James was safe – and his country united against the Catholic cause. Was that what King James and Robert Cecil wanted all along?

Catholics and Protestants still feared one another and did so for a long time to come. But today everybody enjoys the fireworks on 5th November.

Timeline

1534   Henry VIII named head of the Church of England. England's break from the Pope in Rome complete.

1547   Henry VIII dies. His young son Edward VI becomes King.

1553   Edward dies. His Catholic half-sister Mary becomes queen. She reunites the English church with Rome.

1558   Mary dies. Elizabeth I, a Protestant, becomes queen. The Church of England breaks again from Rome.

1570   Guy Fawkes born in York.

1577   Sir Francis Drake begins his 3-year voyage around the world.

1579   Guy's father, Edward Fawkes, dies.

1581   Guy's mother, Edith, remarries a Catholic, Denis Bainbridge. The Dutch declare independence from Spain, leading to war.

1587   Mary Queen of Scots beheaded.

1588   The Spanish Armada defeated by the English navy – and bad weather.

1592 Guy goes to fight against the Dutch for Catholic Spain.

1599 The Globe Theatre opens in London. Shakespeare's plays staged there.

1601 Earl of Essex executed for treason.

1603 Elizabeth dies and is succeeded by her cousin James VI of Scotland; now King James I of England.

1604 **April:** Guy Fawkes returns to England to join Catesby's plot. **Autumn:** Parliament not called because of plague in London.

1605 **26th October:** Lord Monteagle gets warning letter and takes it to Cecil. **30th October:** Gunpowder now in cellars beneath Parliament. **4th November:** First, unsuccessful, search of the cellars. **4-5th November:** gunpowder found. Guy arrested. **8th November:** Rest of plotters either shot or captured.

1606 **29-30th January:** Guy and other remaining plotters are executed.

1611 The English Authorised King James version of the Bible published.

Glossary

armada A large group of war ships.

execute To kill someone as a punishment.

fuse The length of material which is burnt to make gunpowder explode.

plague Deadly disease caught by humans from rat-fleas.

plot A secret plan, usually against the law.

rack A machine used to torture people by stretching them.

rebellious Describes someone who seeks to rebel or go against the rulers of a country.

torture Hurting someone, often very badly, in order to punish them or make them admit to a crime.

traitor A person who does something which damages or betrays their country or the rulers of their country.

treachery The actions of a traitor.

treason Trying to damage or overthrow the rulers of a country.

trial A hearing to decide if a person has broken the law.